For Mom and Dad who have always indulged my dino obsession.
And for Uncle Jay and Aunt Kathy for giving me my first dinosaur book.
–M.G.S.

For my family. Thanks for constantly encouraging me to follow my passion,
and for teaching me that "the easy way" is not a path worth taking.
–S.S.

Otter Illustration Books and **Shawna Smyth Studio** with **&MADE Productions**
self-published in Denton, Texas

This book was created in 2015 in Denton, Texas.

The text for this book is set in Libby and Myriad Pro

The hand-illustrated typography is done by Shawna Smyth using pen and ink.

The illustrations are done by Matthew Gordon Sallack using pen and colored with Photoshop CS4

The book is printed by Amazon's Create Space publishing services.

ARRHINOCERATOPS to ZEPHYROSAURUS

a DINO **ABC** BOOK

ILLUSTRATIONS
MATTHEW GORDON SALLACK

CALLIGRAPHY
SHAWNA SMYTH

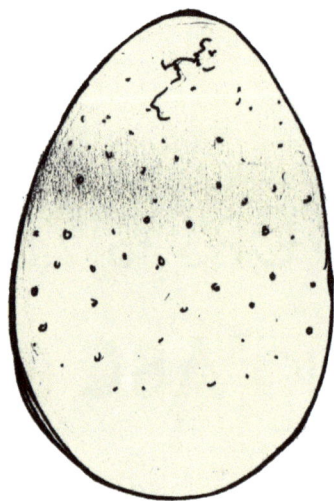

Arrhinoceratops

ah-RHINE-oh-sair-a-tops

Barapasaurus

buh- RAP- puh- SAWR - us

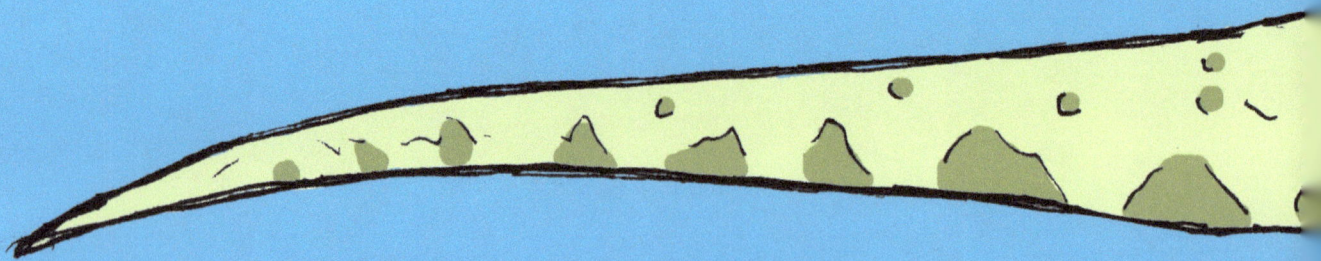

Ceratosaurus

sih-RAT-uh-SAWR-us

Dyoplosaurus

dye-OP-lo-SAWR-us

Elaphrosaurus

ee-LAFF-ruh-SAWR-us

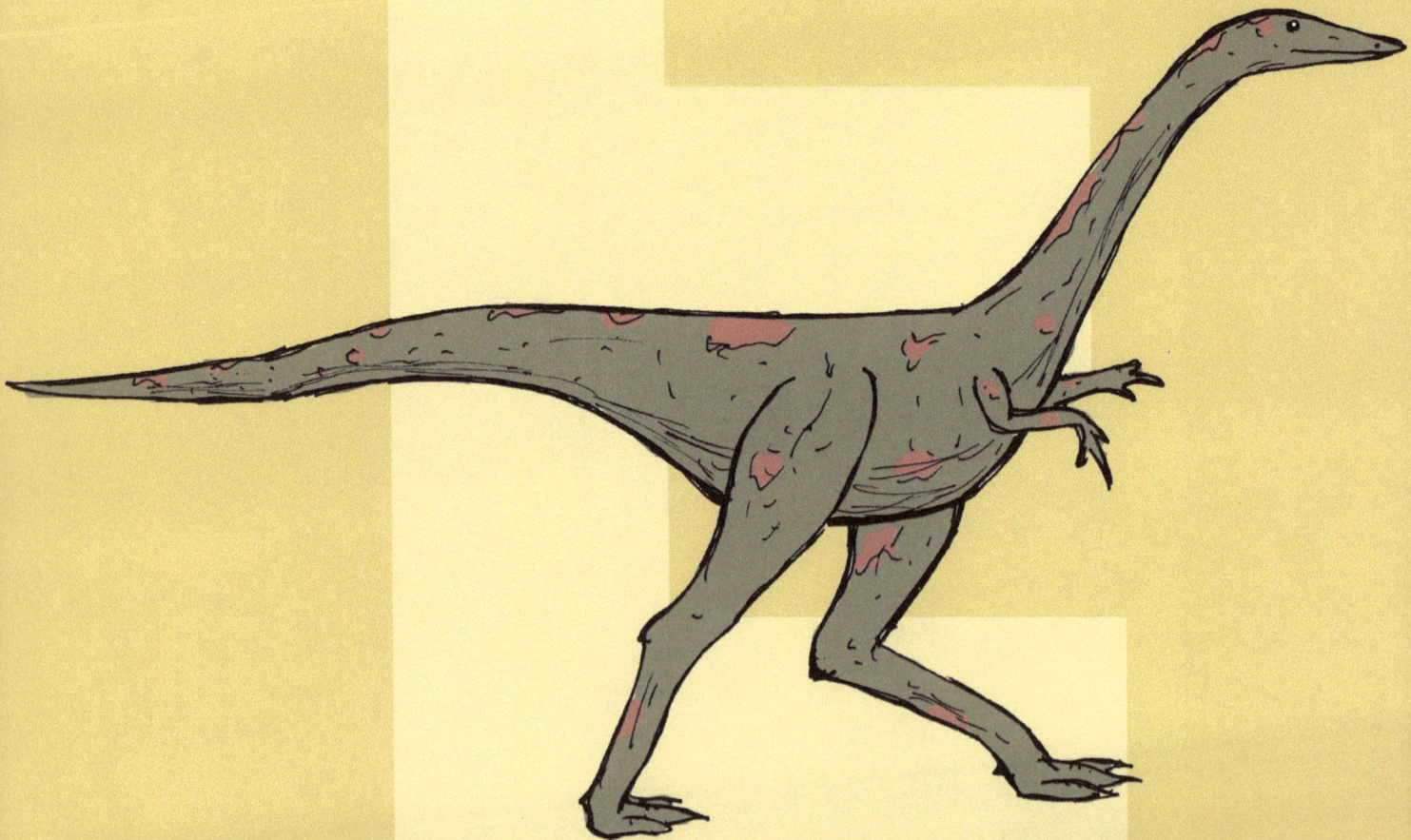

Fabrosaurus

FAB-ruh-SAWR-us

Geranosaurus

jer - AHN - uh - SAWR - us

Hadrosaurus

HAD-ruh-SAWR-us

Indosuchus

in-doe-Sook-us

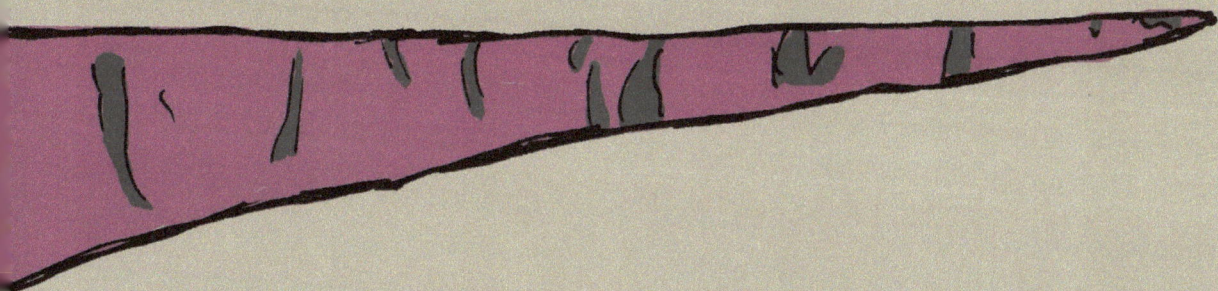

Jaxartosaurus

jax- AR- tuh- SAWR- us

Kentrosaurus

KEN-truh-SAWR-us

Leptoceratops

lep - toe - SAIR - uh - tops

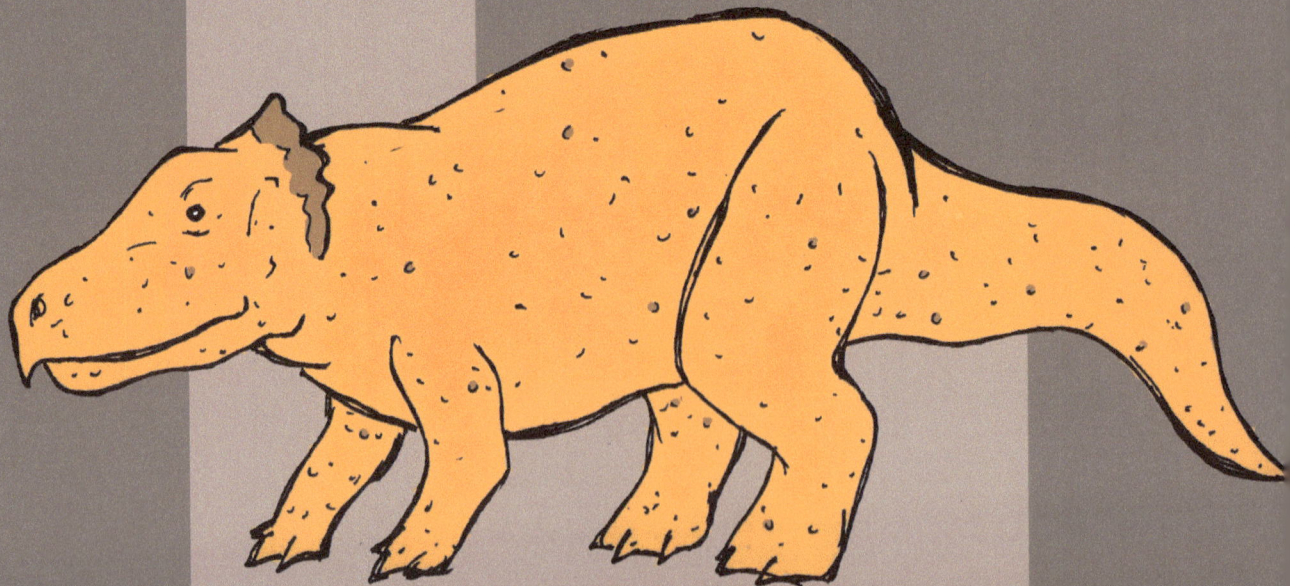

Mamenchisaurus

mah- MEN -chih- SAWR- us

Nodosaurus

no-doe- SAWR -us

Ornithomimus
or-nith -uh-MY- mus

Pachycephalosaurus

pak-ee-SEF-uh-lo-SAWR-us

Quetzalcoatlus

kwet - zuhl - KWAHT - uhl - us

Rhoetosaurus

REE - tuh - SAWR - us

Scelidosaurus

skel - EE - doe - SAWR - us

Torosaurus

TOR - oh- SAWR - us

Utahraptor

Yoo - tah - RAP - tor

Vulcanodon
vul - CAN - uh - don

Wuerhosaurus

WER - oh - SAWR - us

Xuanhanosaurus

zwhan - HAN - oh - SAWR - us

Yaverlandia

yah - ver - LAND - ee - ah

Zephyrosaurus

ZEF-fer-uh-SAWR-us

DRAWING DINOSAURS...

Because no person has seen an actual live dinosaur, we are making a guess every time we draw one. Sure, we have bones and other clues to suggest what they would've looked like. But we have no skin or soft tissue preserved over the millions of years to go by. So who's to say dinosaurs didn't have zig-zag patterns or stripes? Whether they were brown with blue spots, or pink with purple spots...

Not to mention that even though we have fossilized bones of dinosaurs, we rarely find a full, well-preserved skeleton. Paleontologists (scientists who study fossils) have to put together pieces of dinosaur skeletons like a puzzle. Sometimes all we have to go by is a single bone. There is so much that goes into constructing the anatomy of a dinosaur!

After all the information is considered, that's when it's an artist's turn to illustrate his or her interpretation of what the dinosaur would look like as if it lived today. That's what makes drawing dinosaurs so fun. It's a chance for an artist to be creative and use the power of imagination!

So I encourage you to draw your own dinosaurs. Kids and grandkids and nieces and nephews should all draw their own dinosaurs. Make sure to teach them that a drawing is an interpretation of what's already there. And that their interpretation is as valid as anyone else's.

–Matthew and Shawna

ARRHINOCERATOPS– Ceratopsian family. Had a neck shield with one short nose horn and two long horns at the top of the head.

BARAPASAURUS– One of the oldest sauropods. Had a long neck, and a long tail. Up to 25 feet high and 60 feet long.

CERATOSAURUS– Carnivore (meat-eater) that walked on two feet. Had a rather unusual horn at the end of the nose.

DYOPLOSAURUS– Ankylosaur family. Armored with bones along the back for protection and an armored tail with a club at the end that could be used as a weapon.

ELAPHROSAURUS– One of the oldest members of the ornithomimid family. Very fast 2-legged carnivore.

FABROSAURUS– Small 2-legged herbivore (plant-eater.) Grew to be about 1.5 feet high.

GERANOSAURUS– Another small 2-legged herbivore. Had large front teeth and canine teeth.

HADROSAURUS– 2-legged herbivore with a beak and rows of plant-eating teeth in the back of its mouth.

INDOSUCHUS– Resembled Tyrannosaurus, but smaller and with duller teeth.

JAXARTOSAURUS– Had a large helmet-like crust on its skull.   Bipedal ornithischian similar to Hadrosaurus.

KENTROSAURUS– Stegosaur family. Had plates lined down its back and spikes on the tail to defend itself against carnivorous dinosaurs.

LEPTOCERATOPS– Small (2.6 feet high) with a parrot like-beak. Like other ceratopsians, had a neck frill, but had no horns.

MAMENCHISAURUS– One of the longest dinosaurs (69 feet long) with the longest neck (36 feet) of any known dinosaur.

NODOSAURUS– Covered with shell-like bony plates all over its back. Unlike other ankylosaurs, had no club to its tail.

ORNITHOMIMUS– Resembled a modern ostrich. Had a larger brain casing, suggesting it was one of the more intelligent dinosaurs.

PACHYCEPHALOSAURUS– Had a large bony dome at the top of the skull, perhaps used in head-to-head combat like modern rams.

QUETZALCOATLUS– Giant flying reptile (not technically a dinosaur) with a wingspan of 36 feet.

RHOETOSAURUS- Sauropod with a box-shaped head and stocky legs.

SCELIDOSAURUS- Walked on all fours, but would sometimes get up on its hind legs to eat leaves in trees.

TOROSAURUS- Ceratopsian family. Had a nose horn and 2 long horns at the top of its skull. Had the largest skull (9 feet) of any known land animal.

UTAHRAPTOR- A ferocious carnivore with large claws and razor-sharp teeth. Probably hunted in packs, attacking larger animals.

VULCANODON- One of the smaller, more primitive sauropods, measuring 20 feet long.

WUERHOSAURUS- Stegosaur family. Walked on all fours and had large bony plates lining the back and sharp spikes at the end of the tail.

XUANHANOSAURUS- Carnivore like Tyrannosaurus, but with longer arms.

YAVERLANDIA- Like Pachysephalosaurus, but much smaller (2.3 feet high.)

ZEPHRYROSAURUS- Small 2-legged herbivore, with ridged teeth for chewing plants

REFERENCES
Please check out these other awesome dinosaur books and websites:

"DINOSAURS" by John Woodward and Darren Naish. DK Smithsonian, 2014.
"The Complete Book of Dinosaurs" by Dougal Dixon. Southwater, 2012.
"DINOSAURS: an A-Z Guide" by Michael Benton. Derrydale Books, 1988.
http://www.dinodictionary.com
http://news.discovery.com/animals/dinosaurs

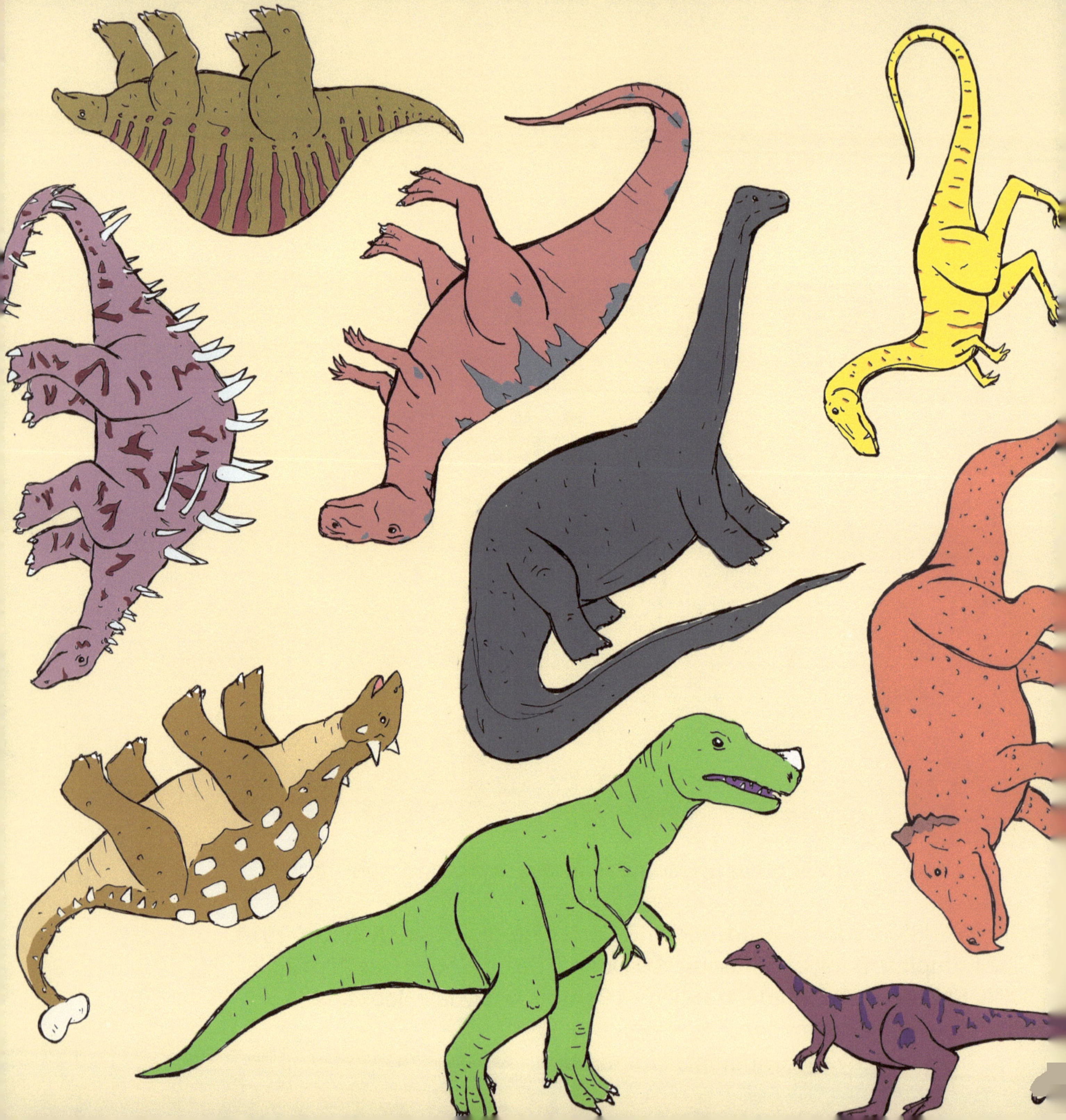